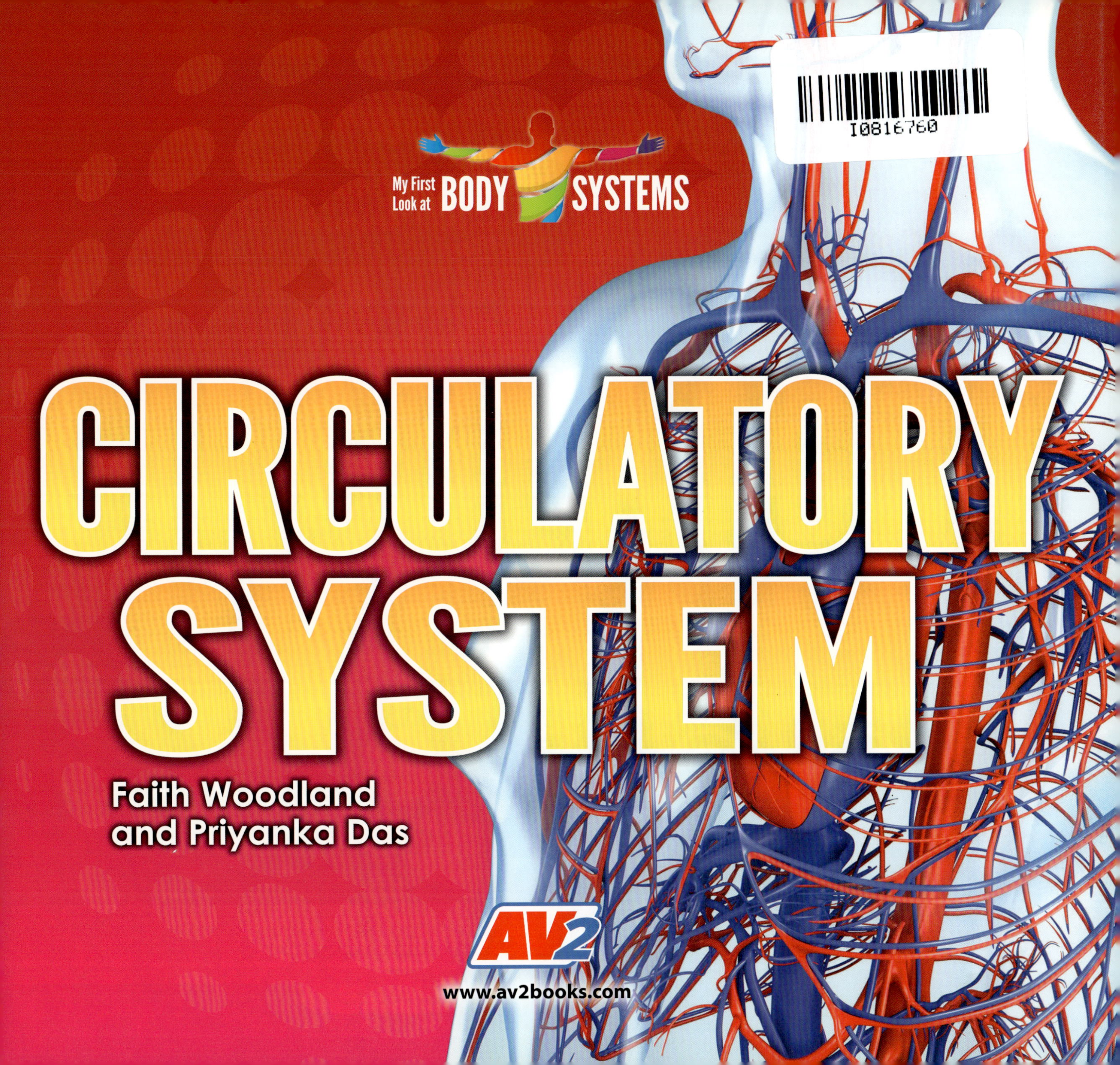
I0816760
My First Look at BODY SYSTEMS
CIRCULATORY SYSTEM
Faith Woodland and Priyanka Das
AV2
www.av2books.com

Step 1
Go to **www.av2books.com**

Step 2
Enter this unique code
NHMTX6VAB

Step 3
Explore your interactive eBook!

My First Look at BODY SYSTEMS

CIRCULATORY SYSTEM

Start!

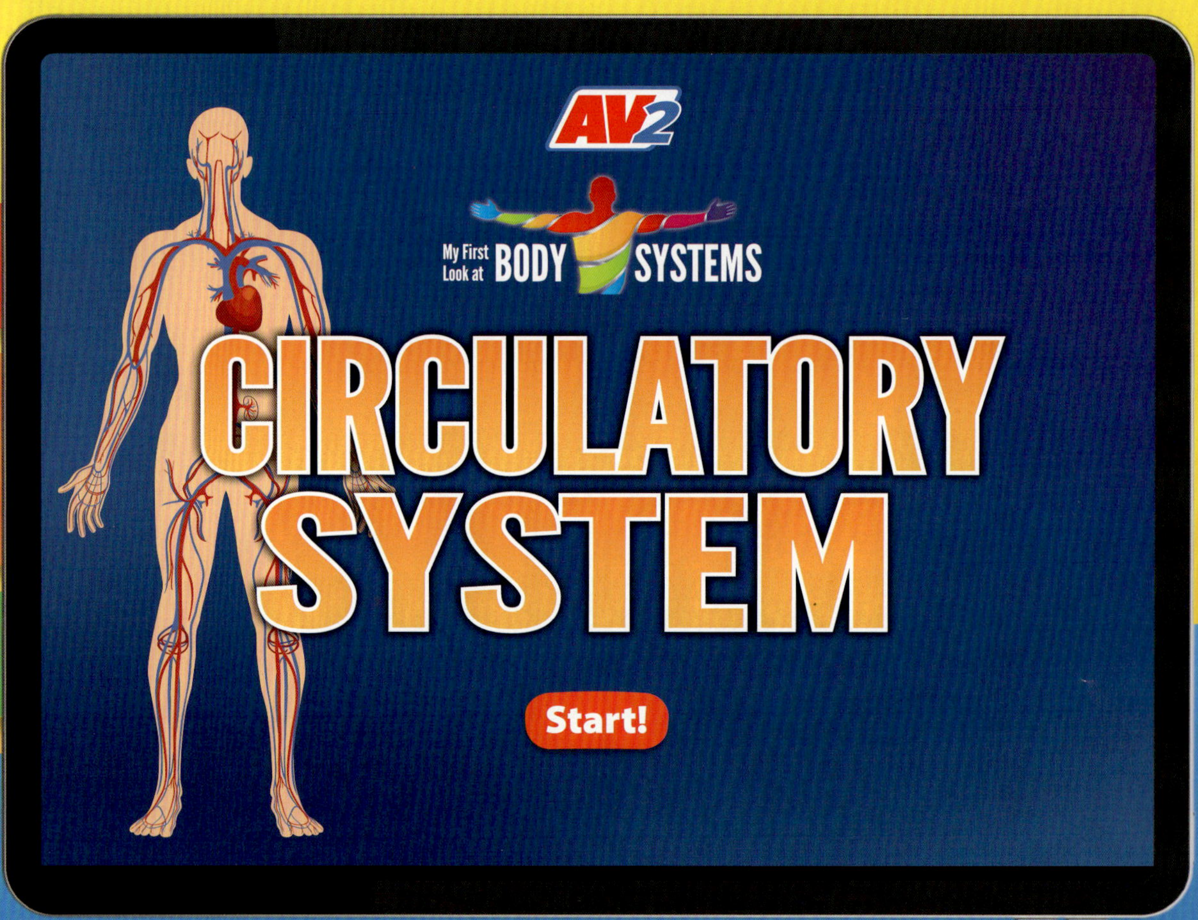

AV2 is optimized for use on any device

Your interactive eBook comes with...

Audio
Listen to the entire book read aloud

Videos
Watch informative video clips

Weblinks
Gain additional information for research

Try This!
Complete activities and hands-on experiments

Key Words
Study vocabulary, and complete a matching word activity

Quizzes
Test your knowledge

Slideshows
View images and captions

View new titles and product videos at www.av2books.com

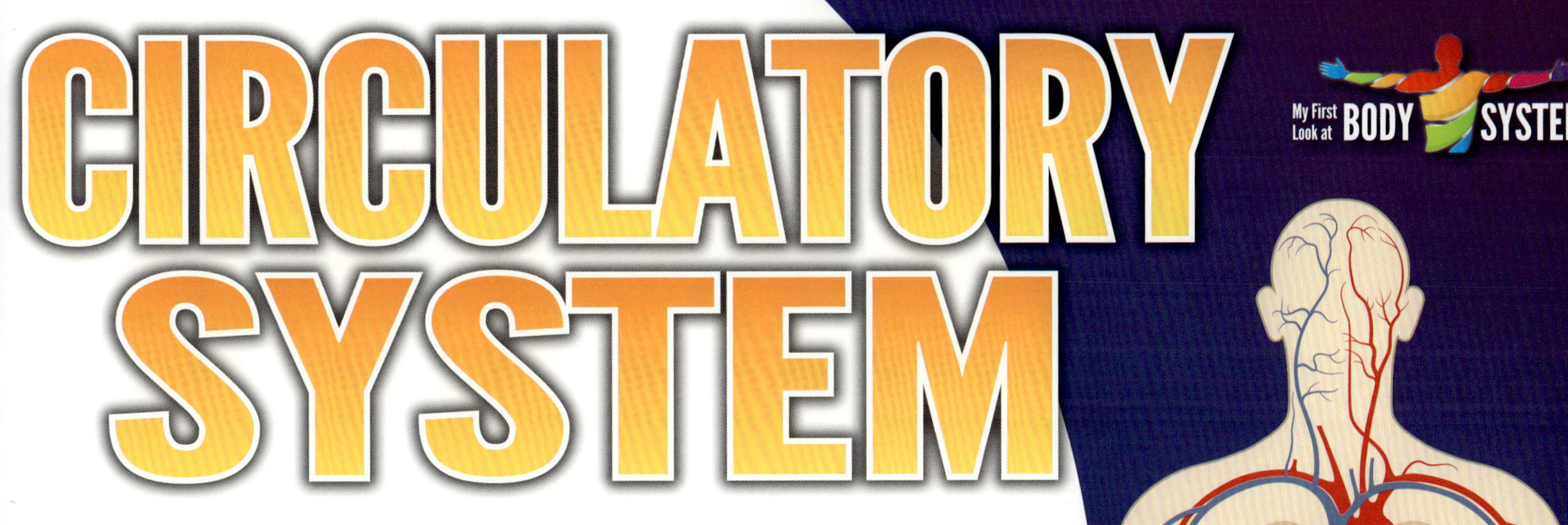

My First Look at BODY SYSTEMS

CONTENTS

Circulatory System

One way to think about the circulatory system is to picture the roads in a city. Highways and smaller roads connect to each other. They transport people and goods through the whole city.

The circulatory system works in a similar way. **It is made up of the heart and blood vessels.** The heart is an organ. It pumps blood. Blood vessels are tubes that carry the blood throughout the body.

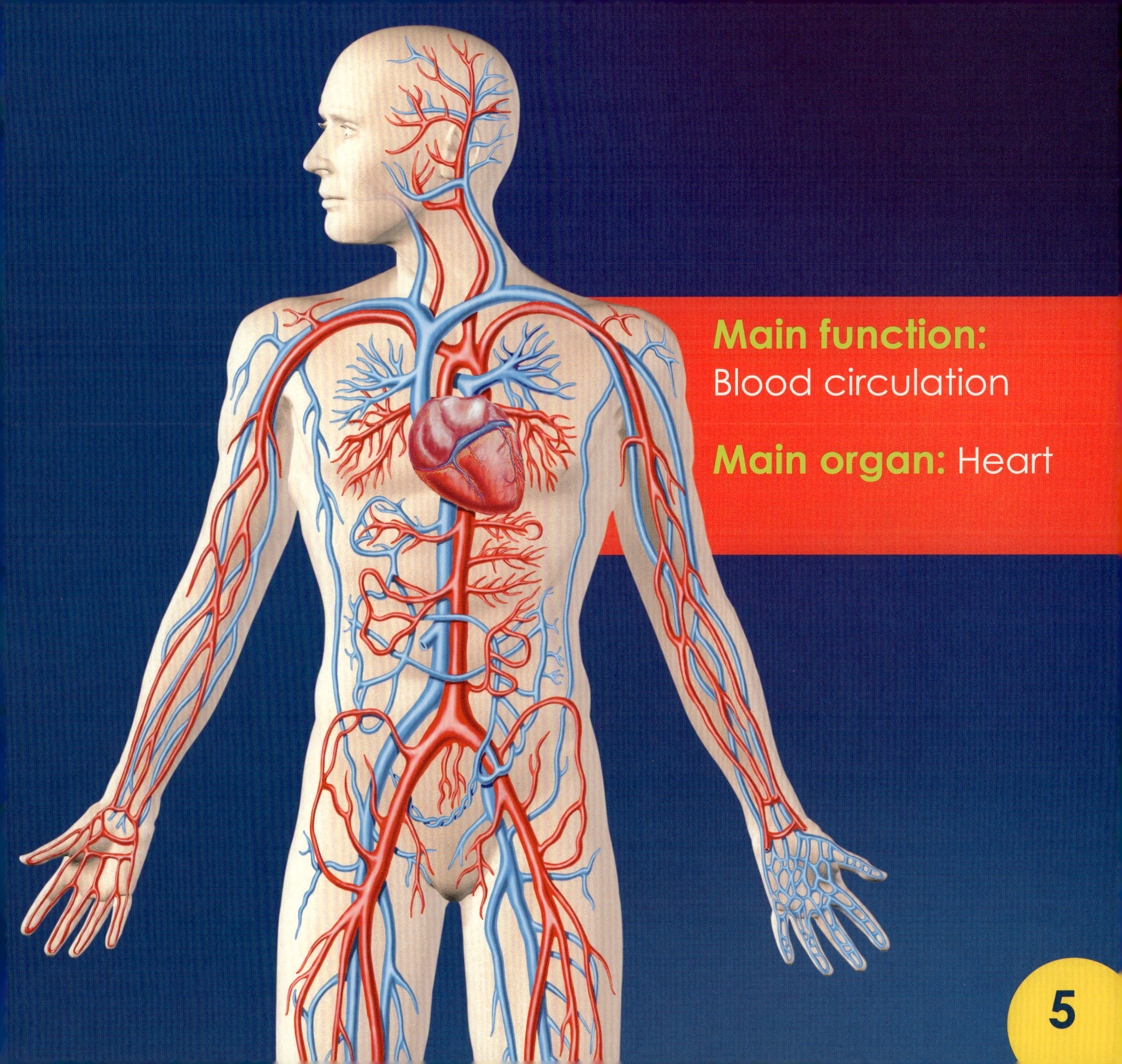
Main function:
Blood circulation
Main organ: Heart

All about Blood

Blood is a very important fluid. **It is needed to keep people alive.** The human body is made up of many, many tiny parts called cells. These cells need oxygen for energy.

Blood carries oxygen and nutrients to every cell in the body. Then, it helps remove the waste produced by the cells. The body also uses blood to help fight infections.

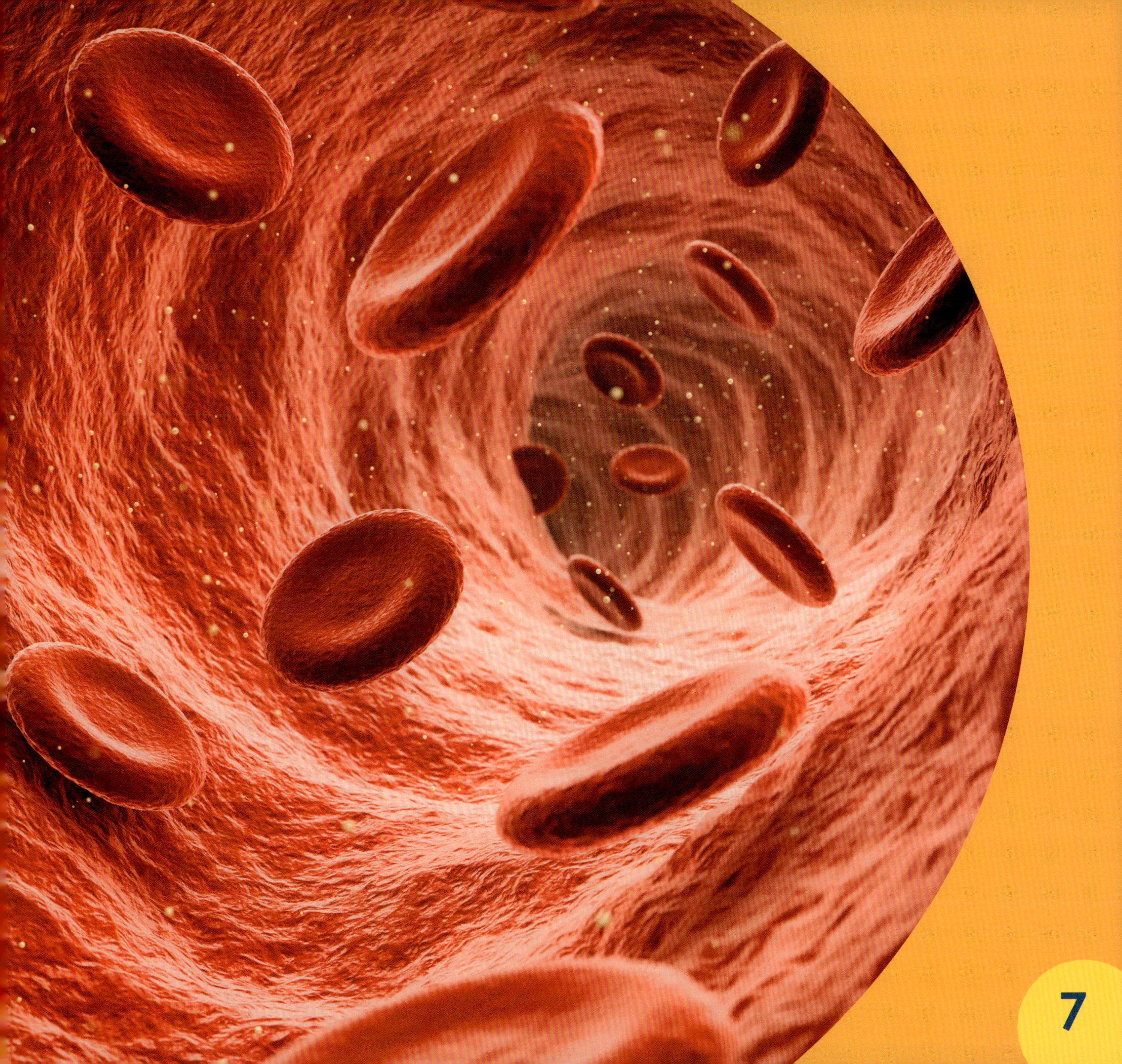

Parts of the Circulatory System

The heart pumps blood to the rest of the body through the blood vessels. **There are three main types of blood vessels.** They are the arteries, veins, and capillaries.

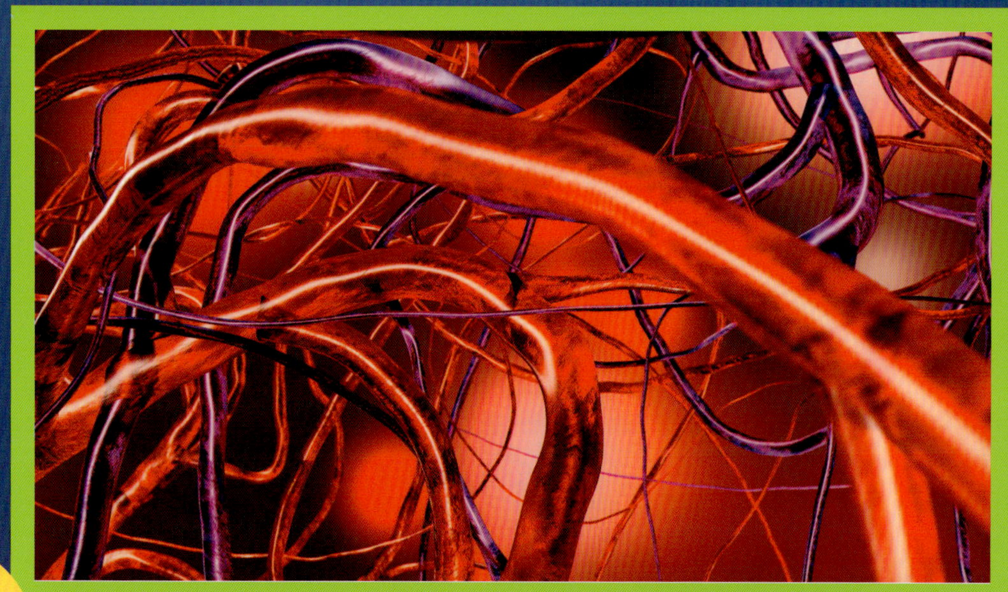

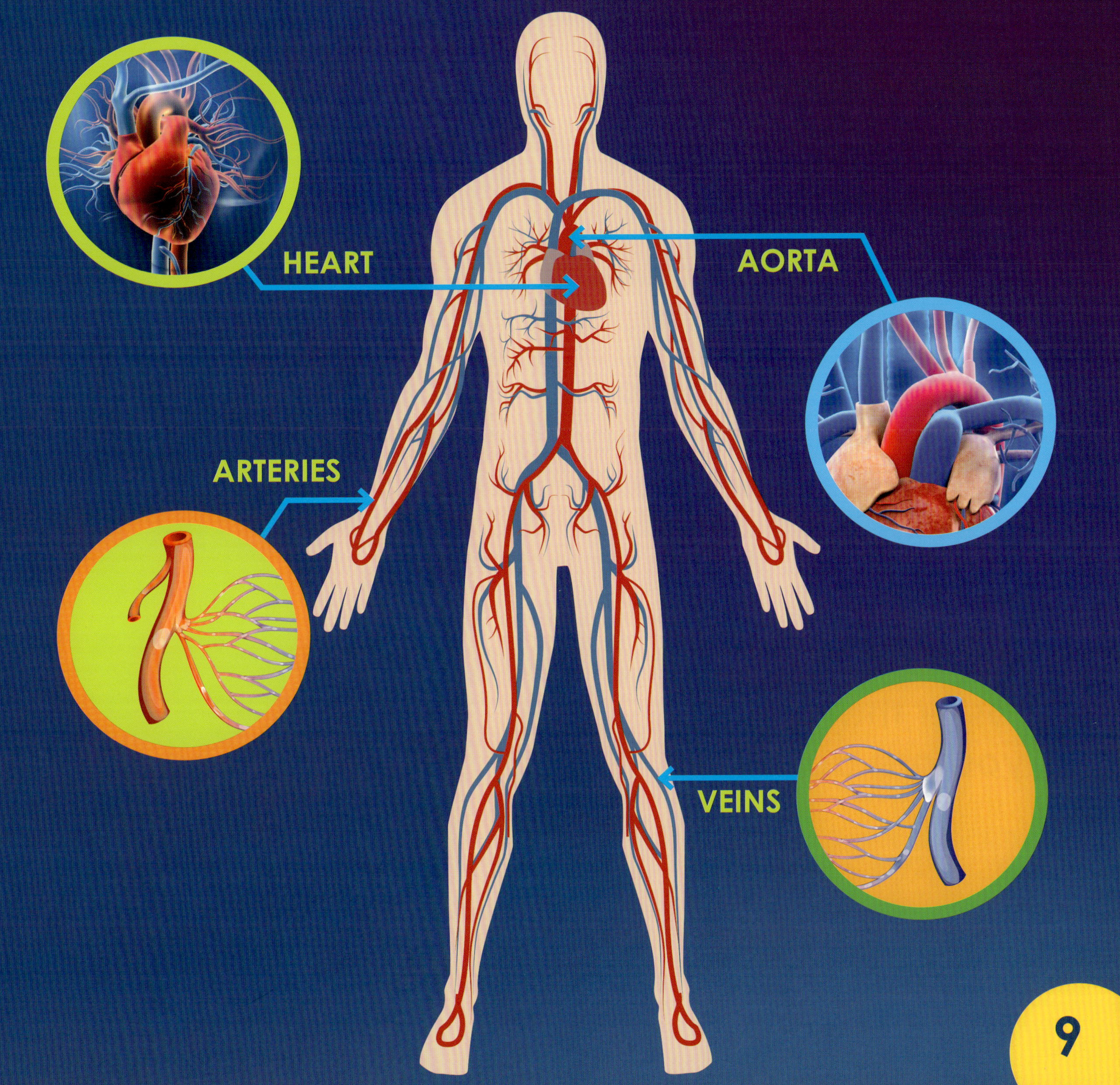
HEART
AORTA
ARTERIES
VEINS

Heart

The heart is the main organ of the circulatory system. It is about the size of a fist. **The heart works with the lungs to provide the body with oxygen-rich blood.**

When your heart pumps blood through your body, you can feel a heartbeat. **The heart beats faster than average when you are active.** It beats more slowly when you are resting.

The heart takes less than **60 seconds** to pump blood to **every cell** in the body.

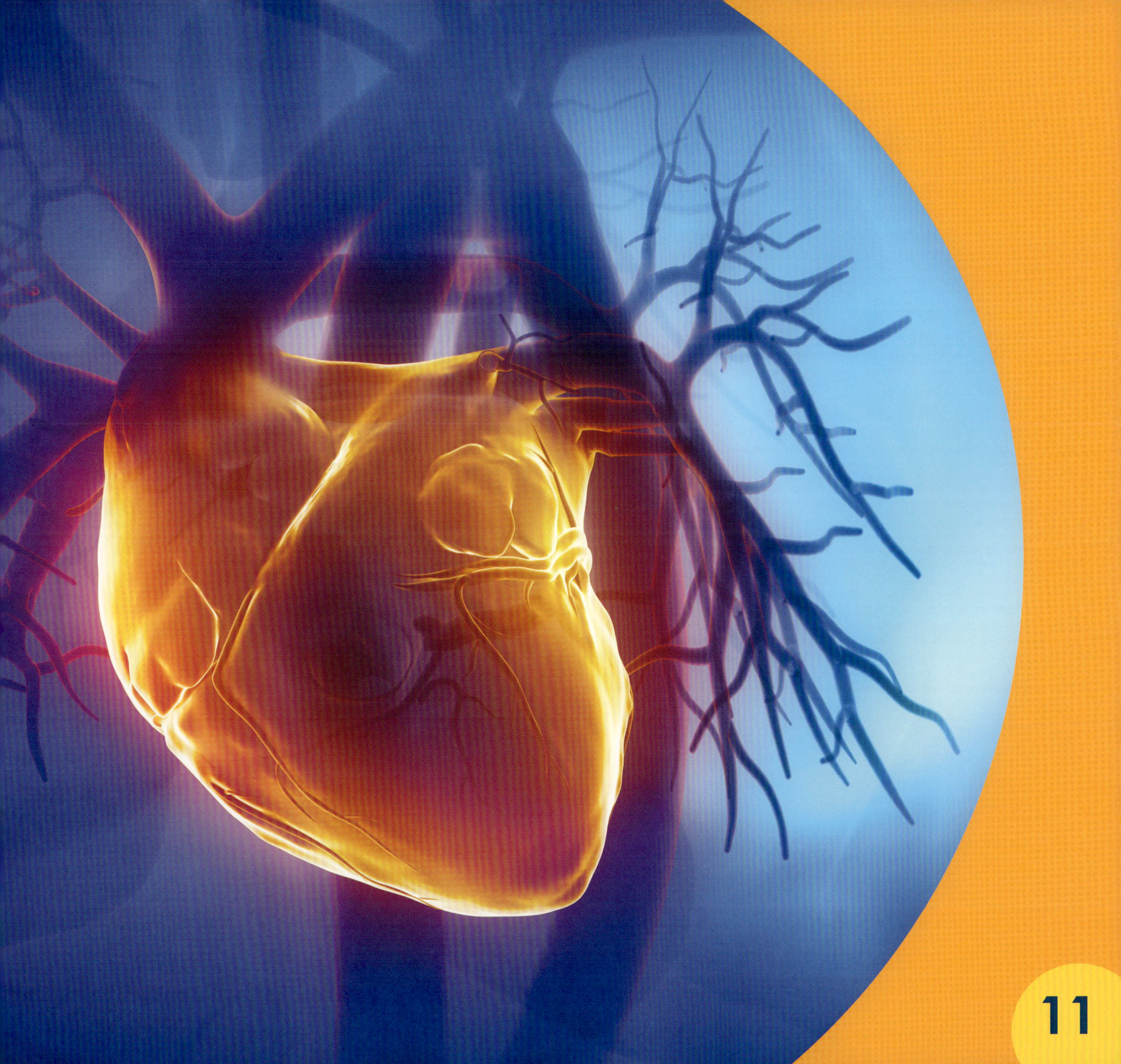

Arteries

Arteries are blood vessels that carry blood away from the heart. Blood that is in the arteries is almost always full of oxygen.

Arteries have thick, muscular walls. **The largest artery is the aorta.** It connects directly to the heart. The aorta branches into smaller arteries. The tiniest blood vessels are called capillaries. Capillaries connect arteries to veins.

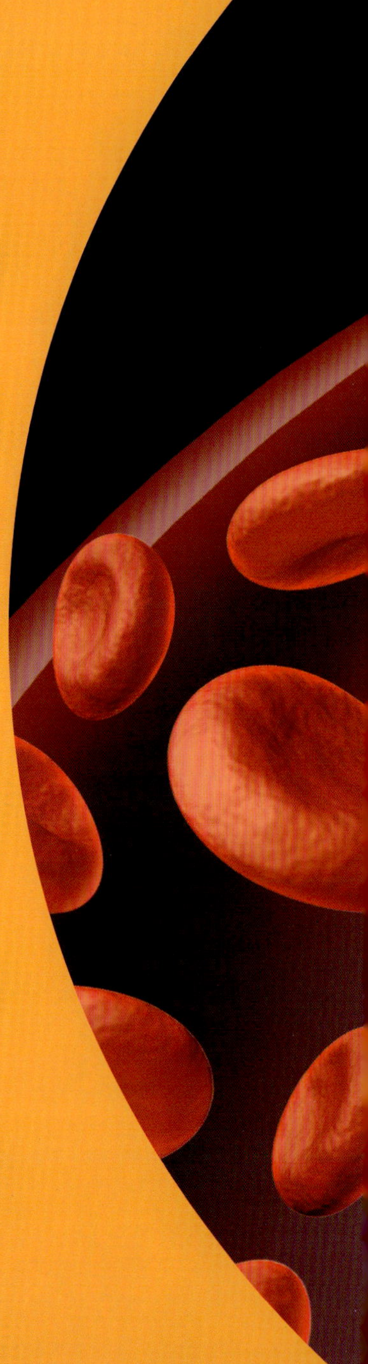

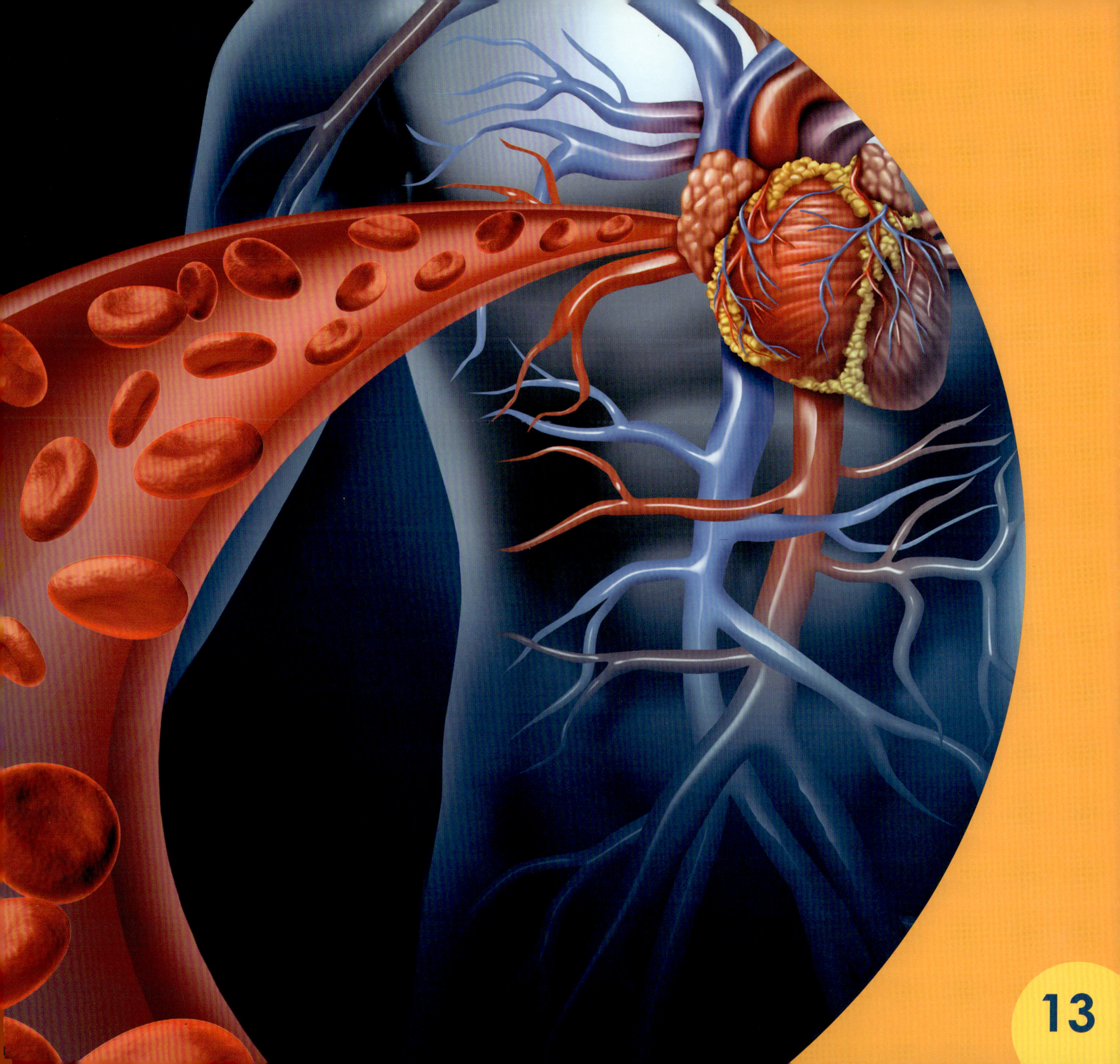

Veins

Veins are blood vessels that carry blood back to the heart. They are thinner than arteries. The blood in veins usually does not have much oxygen.

From outside the body, veins may look blue. However, the blood in them is actually dark red. The blood in veins is a darker color than the blood in arteries.

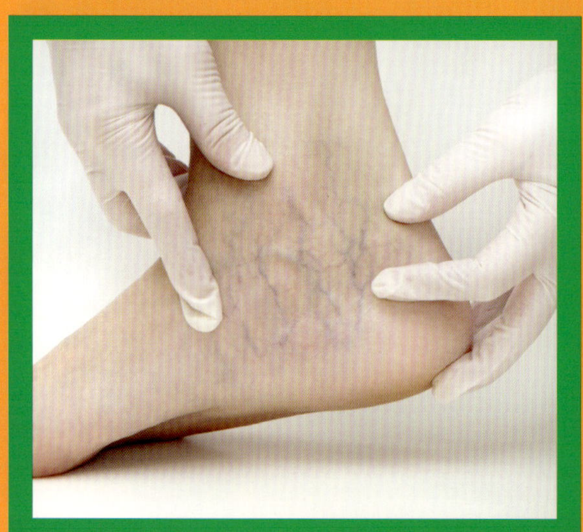

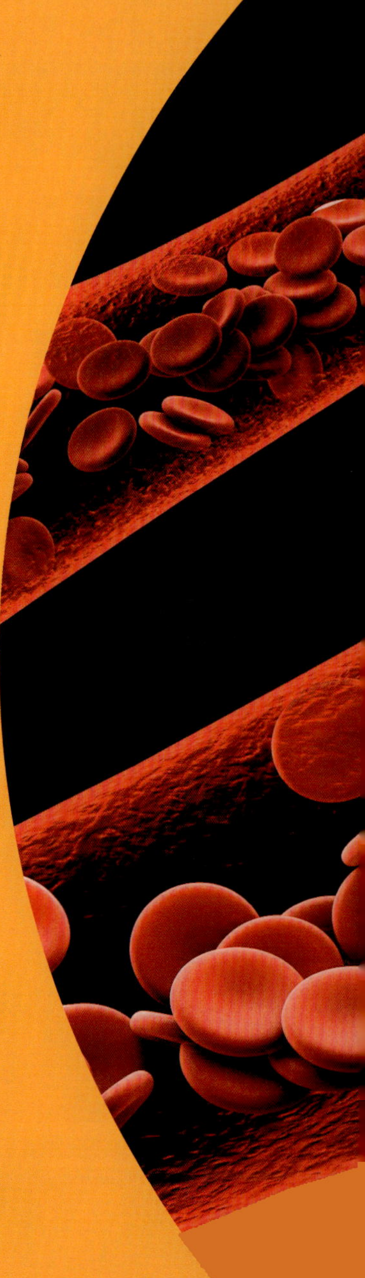

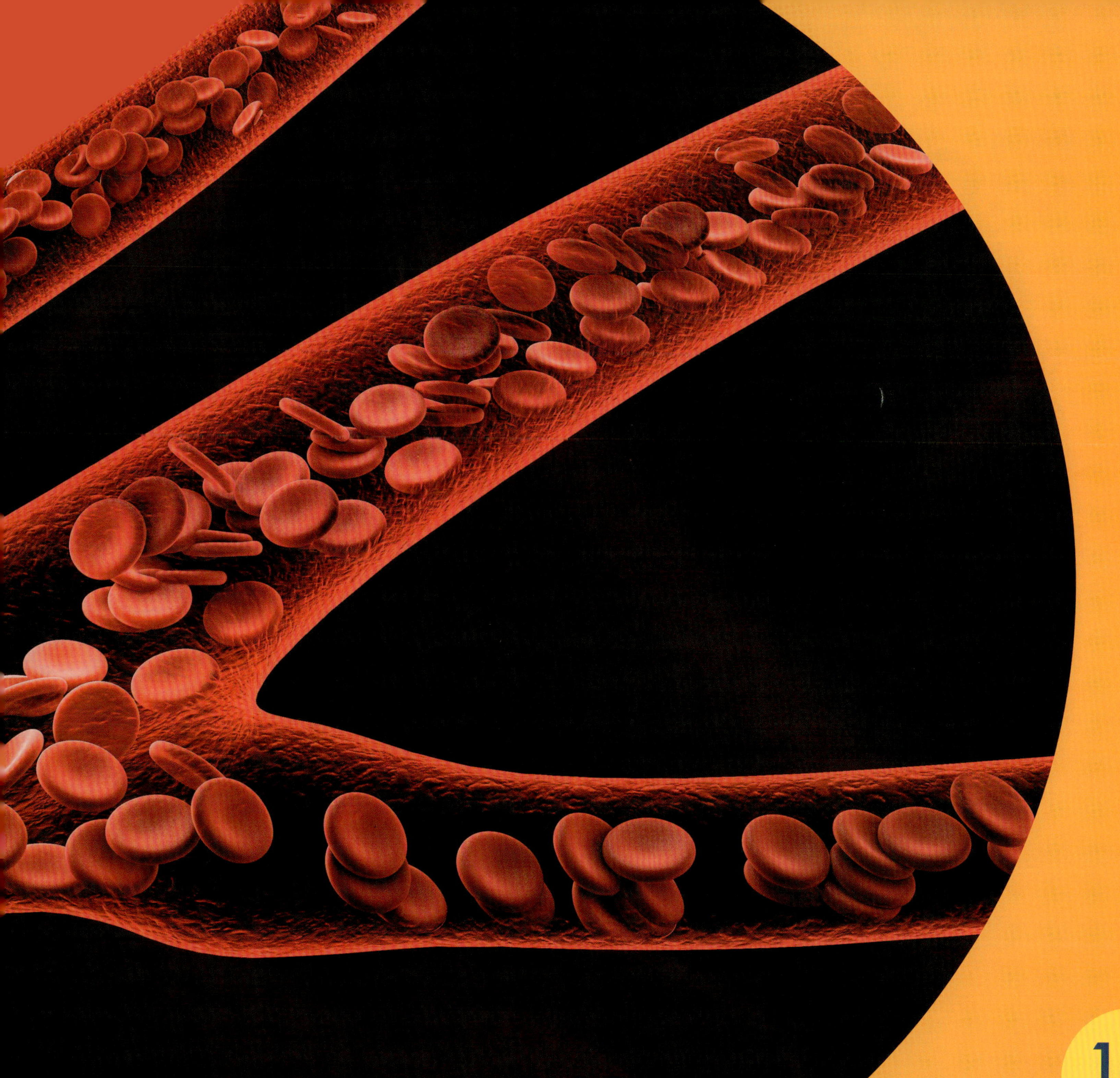

Blood

Blood is bright red when it is full of oxygen. **As blood circulates and oxygen is removed, the blood becomes darker.** People have slightly different types of blood. There are eight main blood types.

Many people are blood donors. This means that they give small amounts of their blood to be stored in blood banks. The blood is used in hospitals to help sick or injured people who share the same blood type.

One blood donation can save up to three lives.

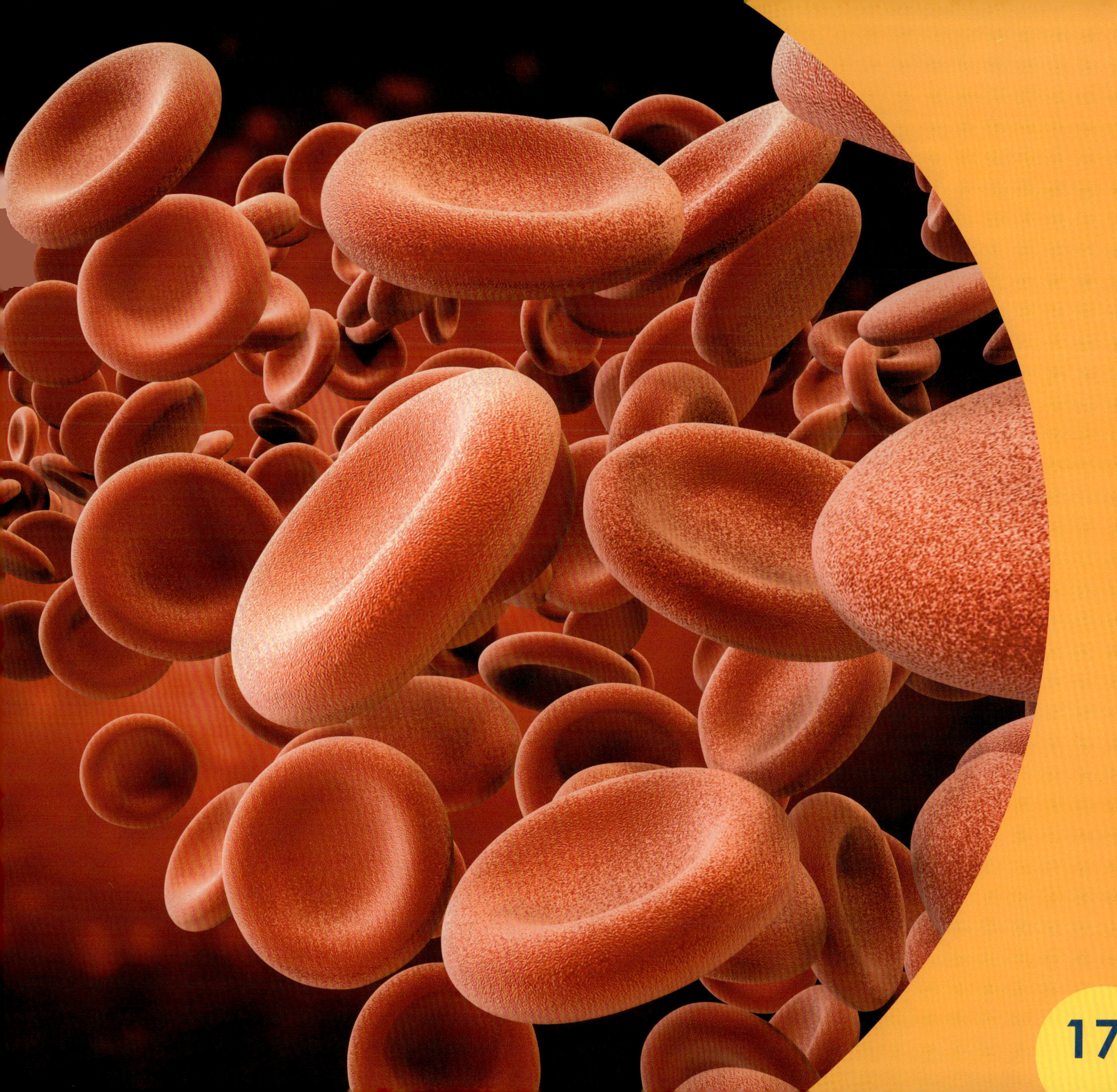

Staying Healthy

Have you ever heard of heart-healthy foods? These are foods that are good for the heart. **Fruits, vegetables, whole grains, and fish are heart-healthy foods**. Foods such as fries, burgers, and pizza are bad for the heart. They can clog the arteries, making it harder for blood to flow.

When you run and jump, your heart beats fast. More blood is pumped through your body. **Exercise helps keep the heart muscle fit.**

Career Spotlight

Paramedics are the first people to respond in a medical emergency. For example, paramedics are called if someone has a heart attack.

A heart attack is when blood stops flowing to part of the heart. This happens when an artery that supplies blood to the heart is blocked. **Paramedics may perform a procedure called CPR on someone who is having a heart attack.** They then drive the person to the hospital in an ambulance.

Cleveland Clinic is the best hospital in the United States for treating heart problems.

PARAMEDIC

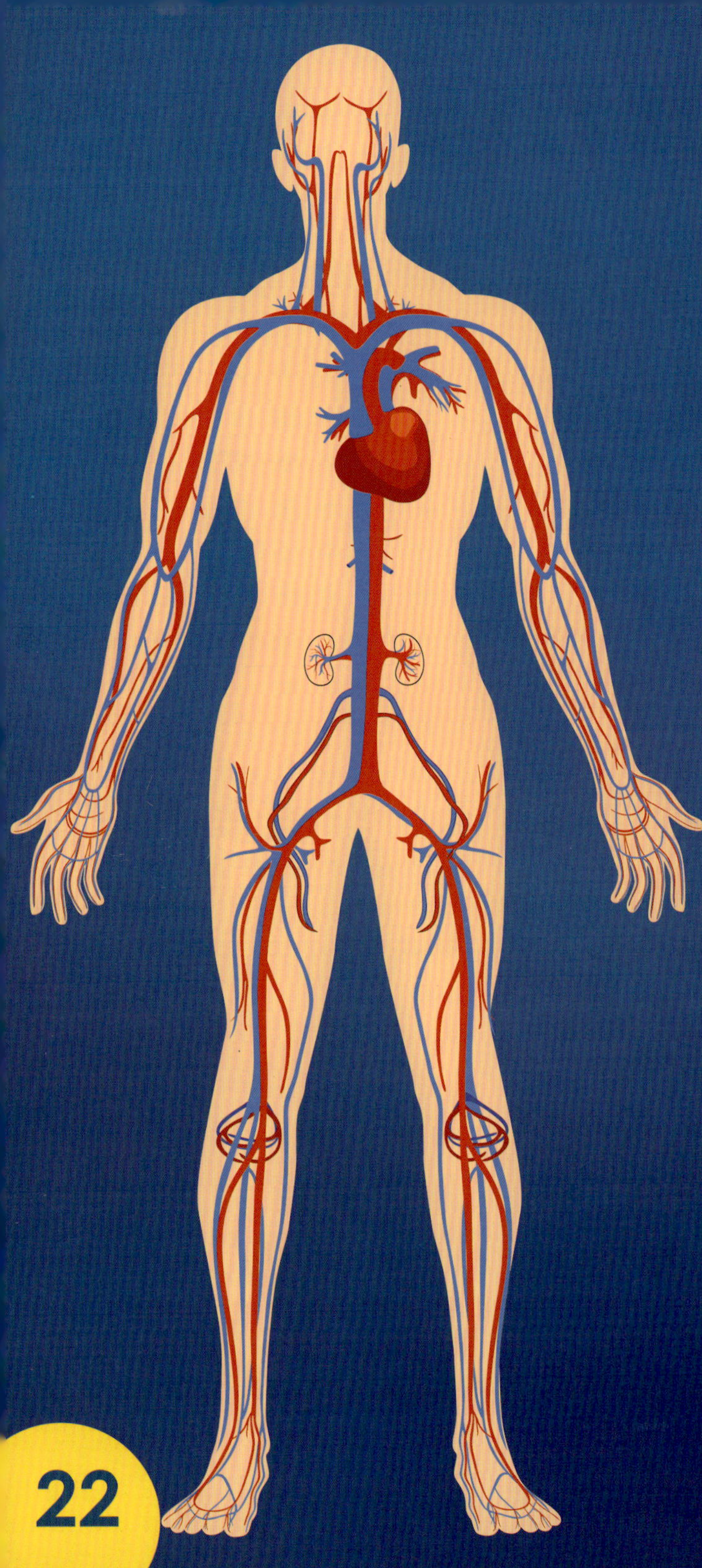

Circulatory System Quiz

Can you name the parts of the circulatory system shown in these pictures?

A] Heart

B] Blood vessels

C] Blood

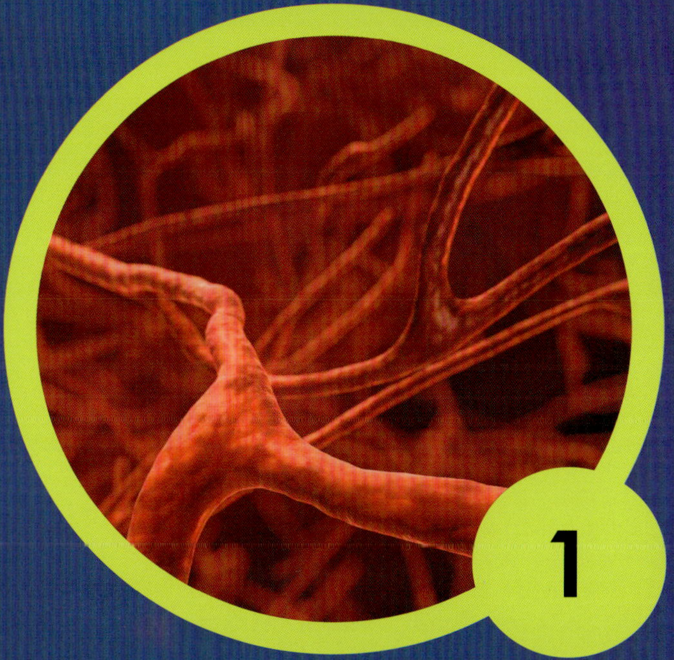

1

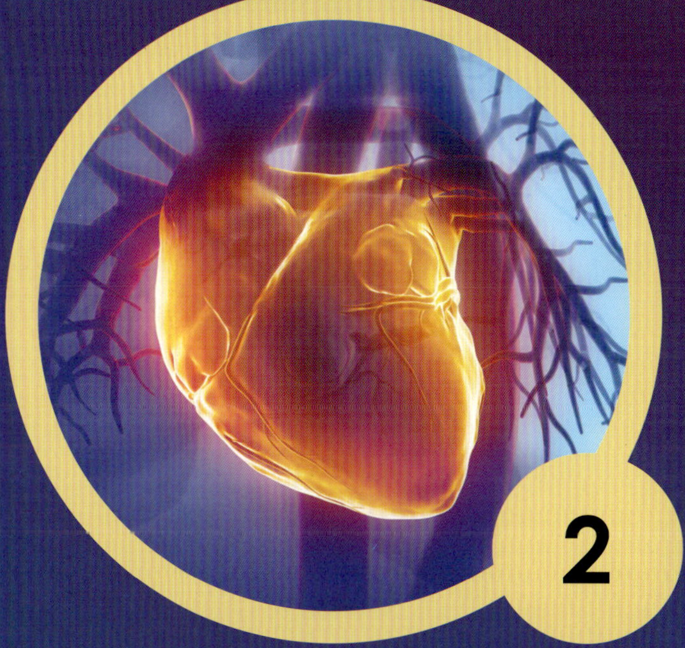

2

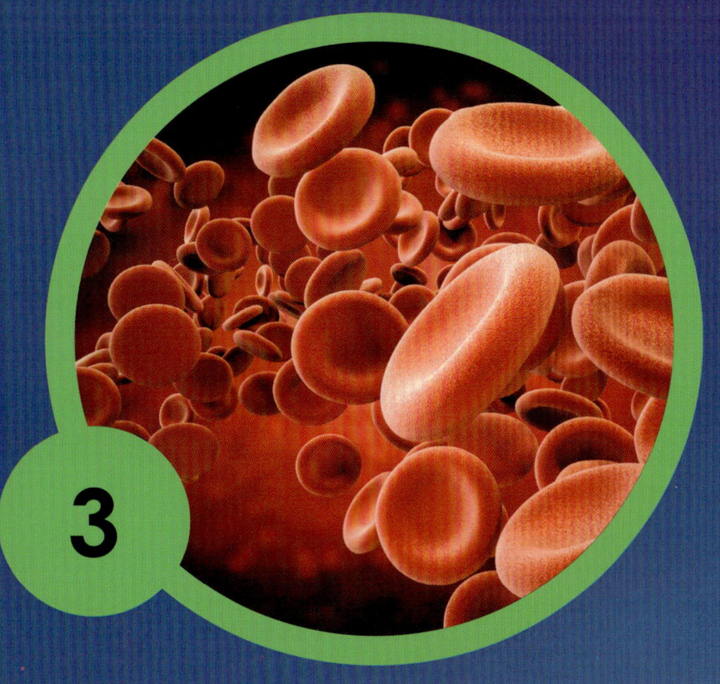

Answers
1] B
2] A
3] C

KEY WORDS

Research has shown that as much as 65 percent of all written material published in English is made up of 300 words. These 300 words cannot be taught using pictures or learned by sounding them out. They must be recognized by sight. This book contains 88 common sight words to help young readers improve their reading fluency and comprehension. This book also teaches young readers several important content words, such as proper nouns. These words are paired with pictures to aid in learning and improve understanding.

Page	Sight Words First Appearance
4	a, about, an, and, are, carry, city, each, goods, in, is, it, made, of, one, other, people, picture, that, the, they, think, through, to, up, way, works
6	all, also, by, every, for, helps, important, keep, many, need, parts, then, these, uses, very
8	there, three
10	can, more, seconds, takes, than, when, with, you, your
12	almost, always, away, from, have, into
14	back, does, look, may, much, not, them
16	as, be, different, give, lives, means, or, same, small, their, this, who
18	foods, such
19	run
20	example, first, has, if, on, states, stops

Page	Content Words First Appearance
4	blood, blood vessels, body, circulatory system, heart, highways, organ, roads, tubes
5	circulation, function
6	cells, energy, fluid, infections, oxygen, nutrients, waste
8	arteries, capillaries, types, veins
9	aorta
10	fist, heartbeat, lungs
12	walls
14	blue, color, red
16	blood banks, donation, donors, hospitals
18	blueberries, burgers, fish, fries, fruits, oatmeal, pizza, spinach, tuna, vegetables, whole grains
19	cycling, exercise, muscle, rope, running, swimming
20	ambulance, Cleveland Clinic, CPR, emergency, heart attack, paramedics, problems, procedure, United States

Published by AV2
14 Penn Plaza, 9th Floor New York, NY 10122
Website: www.av2books.com

Library of Congress Cataloging-in-Publication Data

Names: Woodland, Faith, author. | Das, Priyanka, author.
Title: Circulatory system / Faith Woodland and Priyanka Das.
Description: New York : AV2, [2021] | Series: My first look at body systems | Audience: Grades 2-3
Identifiers: LCCN 2020018390 (print) | LCCN 2020018391 (ebook) | ISBN 9781791118969 (library binding) | ISBN 9781791118976 (paperback) | ISBN 9781791118983 | ISBN 9781791118990
Subjects: LCSH: Cardiovascular system--Juvenile literature.
Classification: LCC QM178 .W66 2021 (print) | LCC QM178 (ebook) | DDC 612.1--dc23
LC record available at https://lccn.loc.gov/2020018390
LC ebook record available at https://lccn.loc.gov/2020018391

Printed in Guangzhou, China
1 2 3 4 5 6 7 8 9 0 24 23 22 21 20

062020
100919

Project Coordinator: Priyanka Das Designer: Jean Faye Marie Rodriguez

Every reasonable effort has been made to trace ownership and to obtain permission to reprint copyright material. The publisher would be pleased to have any errors or omissions brought to its attention so that they may be corrected in subsequent printings.

The publisher acknowledges iStock and Shutterstock as the primary image suppliers for this title.